MW00571568

Fossils Over Time

Q: Are dinosaurs the only fossils that exist?

A: No. Fossils are remains of life long ago. Many plants and animals also lived at the same time as dinosaurs. Their fossils also give us information.

Leaf fossil

Ocean animal fossils

Q: Were all dinosaurs large?

A: No. Both large and small dinosaurs roamed our Earth.

Some dinosaurs were about your size! This dinosaur is a Troodon (TROH oh don). They were about 4 feet tall and weighed about 110 pounds. How does your height and weight compare?

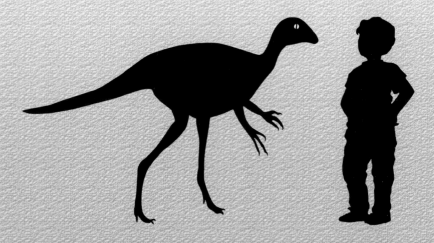

Q: Which dinosaur was the smallest?

A: The Microraptor (my kroh RAP tur) is the smallest known dinosaur. The prefix *micro-* in this dinosaur's name means "small." A microraptor was 15 inches (39 centimeters) long from nose to tail.

Identify some objects that are about 15 inches long.

A beach ball is about 15 inches wide.

Microraptor fossil

Q: Which was the largest dinosaur?

A: The Brachiosaurus (brak ee uh SAWR us) was once thought to be the largest dinosaur. It was about 75 feet (23 meters) long! Scientists have since discovered other dinosaurs that were larger than the Brachiosaurus.

Triceratops ate plants.

Many dinosaurs were bigger than the cars we drive. How many cars long was a Stegosaurus? A Tyrannosaurus rex?

How Many Cars Long?		
Brachiosaurus	75 feet	🚗🚗🚗🚗🚗
Stegosaurus	45 feet	🚗🚗🚗
Triceratops	30 feet	🚗🚗
Tyrannosaurus rex	50 feet	🚗🚗🚗🚗

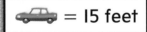 = 15 feet

Q: How many feet tall was a Brachiosaurus?

A: Brachiosaurus was about 50 feet tall. A Brachiosaurus's knee would be about the height of a second grader. About how many giraffes would you have to stack on top of each other to equal the height of a Brachiosaurus?

Second Grader
4 feet
48 inches

Giraffe
16 feet
192 inches

Two-story house
25 feet
300 inches

Brachiosaurus
50 feet
600 inches

Q: What was the weight of a Brachiosaurus?

A: A Brachiosaurus weighed 120,000 pounds.

It would take about 2,000 second graders, or 100 classrooms full of students, to equal the weight of a Brachiosaurus!

Here are some things that are about the same weight as a Brachiosaurus.

400 big
football players

86 cows

120 grizzly bears

Q: Which dinosaur is the most popular?

A: The Tyrannosaurus rex, or T. rex, is probably the most popular dinosaur. It wasn't the fastest or biggest, but it is the most well-known. In fact, the T. rex has been the star of many dinosaur books and movies.

Many museums have the skeleton of a T. rex on display.

Did You Know?

The brain cavity of a T. rex is only big enough to hold 4 cups of water.

Tyrannosaurus rex skeleton

You can create a survey to find out which dinosaurs your classmates like best.

One way to display the results of your survey is in a bar graph.

Which dinosaur from this sample survey is the most well-liked?

Our Favorite Dinosaurs	
Dinosaur	Votes
Brachiosaurus	|||
Microraptor	||||
Troodon	~~||||~~	
Tyrannosaurus rex	~~||||~~ ||

Scientists have discovered about 700 different kinds of dinosaurs. Fossils teach scientists a lot. A fossil can reveal a dinosaur's size and shape, and even what it ate! A fossil can even reveal what the land was like.

By the time you read this book, new fossils may have been discovered. We still have a lot to learn about dinosaurs!

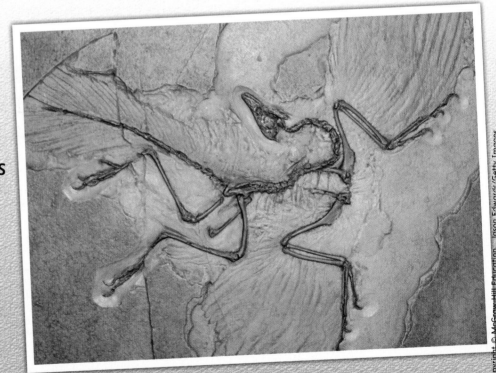